HOW TO BE A STUDENT AND NOT DESTROY PLANET EARTH

Jon Clift & Amanda Cuthbert

> Never doubt that a small group of thoughtful committed citizens can change the world.... indeed, it is the only thing that ever has.

Margaret Mead

First published in 2009 by
Green Books , Foxhole, Dartington, Totnes Devon TQ9 6EB, UK

www.greenbooks.co.uk

Design by
russellhancock.com

Text and covers printed on 100% recycled paper
by Latimer Trend, Plympton, Plymouth, UK.

ISBN 978 1 900322 71 3

Bulk prices available on request –
see www.greenbooksguides.co.uk

HOW TO
BE A
STUDENT
AND NOT
DESTROY
PLANET
EARTH

CONTENTS

HOW TO AND NOT DESTROY PLANET EARTH

01
WHAT IS
HAPPENING
TO OUR
WORLD?

> This is the moment when we must come together to save this planet. Let us resolve that we will not leave our children a world where the oceans rise and famine spreads and terrible storms devastate our lands.

Barack Obama

THE CLIMATE OF OUR WORLD is changing as a result of what we, the humans on this planet, are doing. It's the single biggest problem ever to face humankind, because if we carry on behaving as we're doing now, we face large rises in both global temperatures and sea levels, which will have disastrous consequences.

There has been a lot of talk but little action over climate change. But new initiatives such as the 10:10 campaign and the film *The Age of Stupid* are on the increase, with people who are tired of inaction campaigning for change at grassroots level.

UK UNIVERSITY AND HIGHER EDUCATION INSTITUTIONS EMIT 3.2 MILLION TONNES OF CO_2 A YEAR.

Climate change is a massive challenge, but there's plenty we can all do to slow it down if we act now. There are over 250,000 new students in the UK each year – all of whom will be in the front line as the effects of climate change start to bite. It is your future, and you have the power to act both individually and as a body. If each student did just one thing to reduce CO_2 emissions, it would make a big difference. Get together, start now – don't wait for someone else – do something!

ACT NOW!

WHAT CAUSES CLIMATE CHANGE?

CARBON DIOXIDE

As we all now know, the main cause of climate change is carbon dioxide (CO_2), a gas that is produced whenever we burn fossil fuels such as oil (in the form of petrol or diesel), gas or coal. We use these fuels to give us the energy that we use every day: power stations burn fuel to make electricity for our homes and for industry. We burn fossil fuels to move our cars, buses, trains and planes, to warm and light our houses, and to manufacture our goods – the list is virtually endless.

When fossil fuels are burned, CO_2 goes into the atmosphere. And, as we burn more and more fossil fuels to feed our ever-increasing demand for energy, the amount of CO_2 gas in the atmosphere increases.

We're also using ever-increasing amounts of water, which requires large amounts of energy to clean and to pump to our homes – and even more energy when it is treated once it has reached our sewers.

CO_2 has always been in the atmosphere, trapping just enough heat for life on earth. But excess amounts of CO_2 are upsetting this delicate balance, and not enough heat can escape back into space. Consequently the temperature of our world is rising, and will continue to do so unless we do something about it.

FACT:

THE LESS ENERGY AND WATER YOU USE, THE LESS CO_2 IS RELEASED.

METHANE

Methane is an extremely powerful greenhouse gas – over 20 times more potent than CO_2 – and because of our activities it is now being released into the atmosphere in vast quantities from several sources:

- **Landfill sites** – organic waste, such as leftover food, decomposes when thrown into a landfill site and produces methane.

- **Frozen tundra** – with temperatures rising, many areas of the world which were permanently frozen are now starting to thaw. Huge areas of frozen wetlands which have methane stored away in the permafrost are now thawing, releasing the methane into the atmosphere in ever-increasing amounts.

- **Sea beds** – methane hydrate is a 'frozen' structure of methane and water which is locked into many sea beds at depths greater than 350 metres by the combination of low temperature and high pressure. The rise in seawater temperatures is causing this hydrate to break down, releasing the methane gas.

WHAT'S THE PROOF?

This is not just a media scare story – this is real. So much so that since 1988 the United Nations has had a team of about 3,000 scientists from all over the world monitoring what is happening, researching what to do about it and advising the world's governments.

Storms, floods and drought are now a frequent feature of our news bulletins, and the proof that our climate is changing is now slapping us in the face.

• The average surface **temperature** of the earth, together with air and sea temperatures, is rising.

• **Glaciers** are retreating, and in some cases disappearing altogether.

• The **ice caps** at the north and south poles are **melting.**

• **Snow and ice**, which for millennia have covered vast areas of frozen land, are now rapidly melting, and for the first time in our history there is now a navigable passage through the Arctic ice in summer.

• **Storms and floods** are increasing in intensity and ferocity, with disastrous consequences.

• Eleven of the last twelve years have been the **warmest since records began.**

• **Weather patterns** are much less predictable than they used to be.

• The **seasons** are changing: flowers are blooming earlier, and some foreign birds are no longer returning home in the winter.

HOW WILL CLIMATE CHANGE AFFECT US?

THE WORLD'S SCIENTISTS PREDICT that, unless we dramatically reduce our CO_2 emissions now, the temperature of the Earth will spiral out of control – things are changing far more rapidly than were first expected. The pleasant picture of the UK merely being a little warmer and all of us living a comfortable Mediterranean existence is not what is on the cards.

• Summer **temperatures will rise**, becoming life-threatening at times.

• Seriously **heavy rainfall** and consequently extensive flooding will continue.

- **Flooding** and **storm damage** will become more frequent in coastal communities, as sea levels rise and storms increase in ferocity.

- As the ice sheets continue to melt, we will face large **sea-level rises**. If Greenland's ice-sheets completely melt, this alone will raise sea levels by six or seven metres, with huge repercussions for many cities and areas that are near the coast.

- Water supplies will be under duress, with **water shortages** becoming acute in some parts of the world. This lack of water to both drink and grow crops will, combined with the flooding, create food shortages and force people to move. Large-scale migration is expected, placing huge social and political pressure on the host countries.

WHAT CAN WE DO ABOUT IT?

This book offers loads of ideas for action. Most of the world's climate scientists believe we have time to prevent climate change spiralling out of control, but we need to act NOW. If we change our lifestyles, we can dramatically reduce our CO_2 emissions and minimise the impact of climate change.

MANY STUDENTS ARE ALREADY ON THE CASE

- In **Exeter University** staff and students have cut CO_2 emissions by 11% over the past two years.

- In **Nottingham University** students put pressure on the university to 'Go Green', and within four months their Vice-Chancellor employed a full-time Environment Manager, who is doing great things!

- The **University of Manchester** has reduced its carbon footprint by 8.4% in the past two years.

- Fresher students at the **University of Warwick** reduced their energy consumption across their halls of residence by almost 9% – that's over 200 tonnes of CO_2.

People and Planet, the largest student network in the UK, will help you organise events and get action taken at your college or university.
http://peopleandplanet.org

AND SOME GOVERNMENTS ARE ON THE CASE TOO

- In **France** a personal carbon tax has been introduced to encourage people to reduce their carbon footprint. From 2011 people will have to pay a tax on gas, oil and coal. This is designed as an incentive to use less fossil fuels.

- **Denmark, Sweden** and **Finland** have had a similar scheme in place for the past ten years.

- **Germany**, **Denmark** and **Spain** have for many years encouraged households to install renewable energy systems by offering financial incentives.

- The **UK** is just about to bring in a similar scheme.

- In the **UK** the government has recently passed a bill committing the **UK** to reducing its CO_2 emissions by 80% by 2050. BUT we can't wait until 2050 – we have to make big cuts to our CO_2 emissions NOW.

SMALL ACTIONS BY EVERY STUDENT

=

BIG

RESULTS.

WE NEED TO WAKE UP TO the problems and be more efficient in the way we use energy.

TURN IT OFF
It takes the energy from two power stations to keep our TVs and other gadgets on standby – if we turned them off, we'd save all that CO_2.

TURN IT DOWN
We are all using more and more energy: keeping our rooms so hot that we can walk around in short sleeves in the winter, leaving our computers and lights on all the time.

USE LESS
The less energy, fossil fuels and water we use, the less CO_2 is released.

THINK LOCAL
Don't chalk up food miles by buying French beans in January that have been flown in from Kenya – buy locally grown produce.

GET OUT OF THE CAR
Walking is free – leave the car at home for short journeys.

THINK!
Think before you turn on the heating or run water or plan a trip – can you save CO_2?

2.5% OF THE AVERAGE UK UNIVERSITY BUDGET (ABOUT 200 MILLION POUNDS) IS SPENT ON ENERGY.

BONUS PRIZE

SAVE MONEY!

Saving CO_2 and saving money go together:

- Less energy and water use = lower bills.
- Go by bike = no petrol bills or bus fares.
- Buy second-hand = more money for you.

CLIMATE CHANGE TERMS

CARBON FOOTPRINT

This is the measure of the amount of carbon dioxide your activities add to the atmosphere. Surprisingly, many items – from apples to bicycles – can have a carbon footprint too, especially if they have been flown thousands of miles or if fossil fuels have been used in their production. Your purchasing choices can make a big difference to your overall carbon footprint.

CARBON OFFSETTING

Can't we simply pay for somebody to plant a few trees to cancel our CO_2 emissions? Whilst in theory this may seem like a good idea, this process, known as 'carbon offsetting', is unfortunately not the way out of the problem.

The theory of carbon offsetting is based on the concept of allowing CO_2 to be emitted now, and then reducing it at a later date. Carbon offsetting generally involves paying a company either to invest in renewable energy projects that may reduce CO_2 emissions in the future, or to plant trees that will possibly take CO_2 out of the atmosphere at some future date. But the problem of excess CO_2 is here today: we can't afford to wait, and need to work in the present.

The setting up of these 'offsetting' projects creates the perception that we can carry on polluting as we are currently doing, and buy our way out of the problem. It is infinitely preferable to cut emissions in the first place.

'Climate change' and **'global warming'** both refer to the same thing, although 'climate change' is a better description because the warming up of the Earth changes our whole climate – including how much it rains, the strength of the wind, when and how much it snows and the frequency and strength of storms.

EMBODIED ENERGY

This is the entire amount of energy used in the production of a product. For example, for a cotton T-shirt this would be the fuel burned when preparing the land, planting the cotton, producing the fertilisers and pesticides used, as well as the harvesting, transportation processing and production of your favourite shirt.

GREENHOUSE GASES

These are gases such as carbon dioxide and methane, which accumulate in the atmosphere and prevent heat escaping into space – they act like a greenhouse around the Earth and keep us warm, but they are being produced in ever-increasing quantities by our activities, and are overheating the planet.

POSITIVE FEEDBACK

Positive feedback is a situation where global warming increases the speed and intensity of a cycle of events.

For example:
Global warming causes the **melting of permafrost.**

Methane is released, which then **increases global temperature**.

The effect of increased temperature leads to **more permafrost melting.**

More methane is released. This increases in intensity with every cycle.

'Positive' in this context doesn't mean good. Positive feedback is amplifying the rate of climate change faster than originally predicted. Unfortunately positive feedbacks also have a tendency to spiral out of control.

KEEP WARM

HOW TO KEEP WARM AND NOT DESTROY PLANET EARTH

Over 25% of the UK's CO_2 emissions comes from heating and powering our homes – that's more than all our transport.

It's easy to reduce your carbon footprint!

TURN IT DOWN

- Turn down the thermostat controlling the temperature of your room or house. You will either have a single control at a central position such as in the hall, or, if you have electric heaters, the thermostat is probably attached to the heater.

- If you're too hot in your room, then turn the heating down or off rather than opening your window.

- If you are going away, turn down the thermostat – rooms only need to be warm enough to prevent the water pipes freezing. 5°C will prevent pipes bursting in cold weather.

FACT:

JUST LOWERING THE TEMPERATURE OF YOUR THERMOSTAT BY 1°C CAN REDUCE YOUR ENERGY CONSUMPTION BY 10%.

LET THE SUNSHINE IN

- Open the curtains during the day – if the sun is shining on your windows it will heat your room.

KEEP THE HEAT IN

- Draw curtains over the windows at night: they provide insulation and help to keep the heat in the room.

- Tuck curtains behind radiators so that they don't prevent heat getting into the room.

- Three-quarters of all your heat escapes through your walls, roof and windows if they're not insulated or double-glazed. While you can't really do much to stop this, you could raise the issue with your warden if you're living in halls. Point out that by insulating and double-glazing the university can cut its heating bills for the halls by about 75%.

FACT:

PAYING FOR ENERGY MAKES A SIGNIFICANT HOLE IN YOUR STUDENT LOAN, SO IF YOU SHARE A HOUSE OR FLAT AND CUT DOWN ON YOUR ENERGY CONSUMPTION YOU CAN SAVE MONEY AS WELL AS DOING YOUR BIT TO REDUCE CLIMATE CHANGE. AND, EVEN IF YOU LIVE IN HALLS, ALTHOUGH YOU WON'T SAVE ANY MONEY DIRECTLY, THE UNIVERSITY WON'T HAVE SUCH A LARGE ENERGY BILL TO PAY – AND HOPEFULLY THEY'LL SPEND THE SAVING ON YOU!

If you have windows that are not double-glazed you can make very effective temporary double-glazing by using cling film – cutting and placing a piece of film over each window surround, and leaving a gap between the window pane and the cling film. It works remarkably well, will reduce noise coming through the window, and should last a winter.

Move furniture away from any radiators or heaters: let the heat get out and circulate around the room.

Keep your external doors shut and block up any draughts, including through the letterbox. If you are living in a house, then why not buy and fit a draught excluder to your letterbox – they cost a couple of quid and make a big difference. BUT don't block up air vents or grilles in walls – it's really important that there is still sufficient ventilation.

! BEWARE – if you have an open gas fire, a boiler with an open flue, or a solid-fuel fire or heater, make sure you still have sufficient ventilation. Be especially careful in situations such as these, as insufficient ventilation could lead to carbon monoxide poisoning!

CAN YOU TURN THE HEATING DOWN?

03
HEAT
WATER

HOW TO HEAT WATER AND NOT DESTROY PLANET EARTH

Heating water for showers and baths is responsible for 5% of the UK's CO_2 emissions, and about 25% of the average energy bill.

TURN IT DOWN

• Turn down the temperature of your hot water at the central heating boiler, at the immersion tank (if your water is heated by electricity) or on your instant water heater. Don't waste energy heating water only to have to add cold water so that it's not too hot to use! 60°C/140°F should do it.

USE LESS

• Take a quick shower rather than a bath.

• If you have a power shower, remember that in five minutes it can use as much energy as a bath.

• If you must have a bath, then obviously it is much more environmentally friendly to share it with a friend!

WRAP IT UP

- If you're in a house or flat then check to see if your hot water tank in the airing cupboard is insulated. It should be totally surrounded either with foam insulation or an insulating jacket. If it isn't, then contact your landlord. Insulating jackets are very cheap but will save a considerable amount of energy – get one that's at least 75mm thick.

If you're in halls and your hot water is too hot, tell your warden.

USE LIGHT

HOW TO USE LIGHT AND NOT DESTROY PLANET EARTH

£ **10% of most electricity bills goes on lighting.**

TURN IT OFF

It's a no-brainer! If there's nobody in the room, or the room is bright enough without having lights on, switch them off. Get into the habit – it costs nothing and it's easy!

GET A LONG LIFE (BULB)

• When you need to replace a light bulb, get a low-energy bulb: they last about 12 times longer than ordinary bulbs and consume about a fifth of the energy that ordinary bulbs do. One of these bulbs in your desk light, for example, will last you all of your student days and beyond. They come in all shapes and sizes, including spotlights.

• Beware of 'uplighters': many use high-wattage bulbs of 300W or greater – that's the equivalent of over 30 low-energy light bulbs!

USE LESS

Use natural light where possible.

If you're living in halls, then your university could reduce its energy bill quite considerably if low-energy bulbs are used in places such as corridors and toilets, which tend to have lights on for most of the time. If your university is still using old-style bulbs, talk to the warden and point out the savings that could be made.

05 COOK

HOW TO COOK AND NOT DESTROY PLANET EARTH

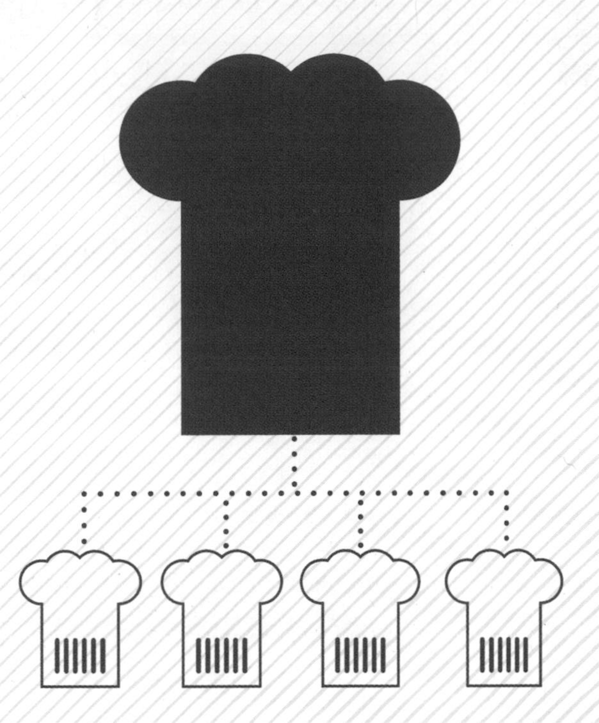

WHILE YOU MAY not spend hours slaving away over a hot stove, it's still an area where you can reduce your energy consumption and cook more efficiently (and more quickly!).

And if you don't or can't cook, then why not start? If you are sharing a kitchen, cooking is a chance to catch up with your mates. It can be a lot healthier for you and is much cheaper than ready-made meals, and it's surprisingly quick once you know what you're doing.

The following simple actions will save you both time and money, and apply whether you're cooking on gas or electricity.

POTS AND PANS

- Put a lid on it!

- Select the correct saucepan size for the heating element and pop a lid on top of the pan when you can – your meal will cook a lot more quickly and you won't be wasting energy.

- When the contents of the pan come to the boil, turn down the heat. You don't need as much heat to keep a pot boiling as you do to get it to the boil, and the contents will cook just as quickly.

- If you're cooking vegetables in saucepans, only put in sufficient water to cover them.

• Why not try cooking with a pressure cooker? They are incredibly effective and reduce cooking times dramatically. Make sure you read the instructions if you don't know how to use one – they are not suitable for cooking some types of food.

• When cooking rice, turn off the heat 5 minutes before the end of the cooking time, keep the lid on and let it finish cooking in its own steam.

• Use a steamer for vegetables – you can cook two or three vegetables on one element or gas ring.

• Make one-pot meals that only need one element or gas ring.

FACT:

IF YOU ALL COOK TOGETHER YOU'LL SAVE ENERGY.

KETTLES

- Use an electric kettle to boil water for cooking.

- When using a kettle don't fill it too full – just put in the amount of water you want, but make sure you cover the element.

- Electric kettles vary hugely in the amount of electricity they consume; when you need to replace yours choose a low-kW one.

- See if your housemates want a cuppa before you make yourself one, and only use as much water as you need. By boiling the kettle less often you will save electricity.

OVENS

• If you're cooking a meal in the oven, don't keep on opening the oven door to see how it's all going – you lose a lot of heat that way and it won't cook any more quickly!

• Use your oven efficiently by filling up as much of the space as possible.

• When using a non-fan-assisted oven, food will cook more quickly on the top shelf – it is much hotter than the bottom.

• Use the grill rather than the oven when you can.

• Make toast in a toaster rather than under the grill if you can.

• Plan ahead: get ready-made meals out of the freezer early enough for them to defrost without using energy. If you are in a hurry, heat or defrost them in a microwave rather than in a conventional oven.

STOP

DO YOU NEED THAT MUCH WATER IN THE KETTLE?

KEEP THINGS COOL

HOW TO KEEP THINGS COOL AND NOT DESTROY PLANET EARTH

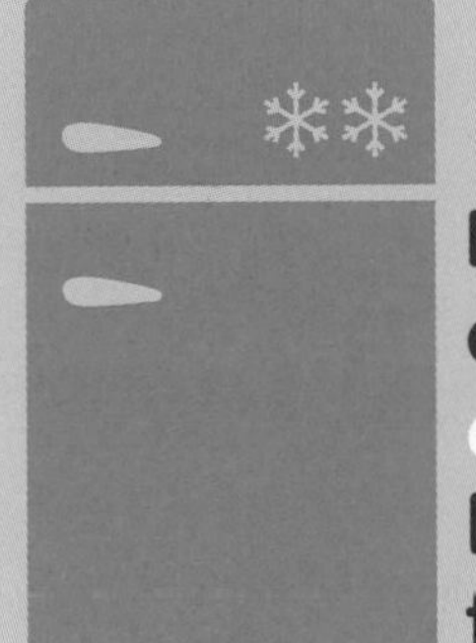

Fridges and freezers account for about one-third of your household's carbon footprint.

FRIDGES AND FREEZERS are never turned off – although they may not appear to use much energy, they are on 24 hours a day.

COOL IT

- Wait until hot food has cooled down before putting it in the fridge.

- Don't keep the door open any longer than necessary.

- Try to keep your fridge and freezer full: they will use less electricity. You can use bubble wrap or newspaper to fill any empty freezer drawers.

- Defrost food by putting it in the fridge the night before you want to use it: this will cool the fridge down and reduce its power consumption.

- Keep fridges and freezers well away from heat sources such as cookers, dishwashers and washing machines if you can.

- If possible, site fridges and freezers out of direct sunlight – your appliance will use more energy trying to keep cool in the sun.

- Keep the metal grids (condenser coils) at the back of fridges and freezers clean and dust-free, and not jammed up against the wall; this allows the air to circulate more easily around them, and makes them more efficient.

DEFROST IT

• Defrost the fridge and freezer regularly and it won't be such an epic task. A defrosted fridge is more energy-efficient.

• If you are all going away during a holiday, just before you leave is a good time to defrost the fridge. The ice inside should never be more than 5mm – greater than this and it becomes inefficient.

HOW TO

defrost your fridge/freezer:

BEWARE

When defrosting, remember to keep all electrical plugs and appliances clear of any water and don't be tempted to attack the ice inside with a knife. You could damage the coiling tubes buried in the ice and ruin the whole fridge.

IS THE FRIDGE/ FREEZER ICED UP?

CLEAN YOURSELF

HOW TO CLEAN YOURSELF AND NOT DESTROY PLANET EARTH

IT MAKES SENSE not to waste water (a lot of energy is involved in treating it and getting it to your taps) – if your house is on a meter then every drop costs you money, and the collection, cleaning and delivery of water is a very energy-intensive process. The less water used, the less CO_2 is released and the more money you save.

YOUR DISHES

BY HAND

- If you only have a few things to wash up, wash them up in a bowl.

- Never leave a tap running – use a bowl to wash vegetables and rinse plates.

- When washing dishes by hand, fill a bowl with warm water and a little detergent, washing the 'cleaner' items first.

- Use cold water for rinsing.

- Get that leaky tap fixed – a tap that drips once a second wastes 33 litres of water a day.

FACT:

THE UNIVERSITY OF BATH CUT ITS WATER USE BY 9% IN A YEAR, SAVING £65,000 IN WATER BILLS.

USING A DISHWASHER

- If you use a dishwasher, wait until it is full before using it. Don't be tempted by the 'half-load' facility, as it is nowhere near as energy-efficient.

- Use the 'economy' or 'eco' programme if your dishwasher has one; it will use less electricity and take less time.

- If you switch off the machine and open the door when the dishwasher enters its 'drying phase' the dishes will dry naturally, saving a considerable amount of energy.

- Switch your dishwasher off completely when it has finished; it is still consuming electricity on standby.

ARE YOU WASHING YOUR DISHES UNDER A RUNNING TAP?

YOUR CLOTHES

WASHING

In an average household the washing machine uses about 230kWh per year, causing about 123kg of CO_2 to be released. It is also a great consumer of water, using 13,500 litres every year. By reducing the number of times you use the washing machine you can dramatically reduce the amount of CO_2 being emitted. It's a simple way to save energy, save CO_2 and save money.

- If you've only got a few dirty clothes, wash them by hand. There is no need to have the water hot – most non-greasy dirt will wash out easily with cold water and detergent, and cold water is fine for rinsing your clothes.

- Wait until you've got a full load before using your washing machine – using the 'half load' programme does not save you half the energy, water or detergent.

- Use environmentally friendly detergent and less of it – or go the whole hog and check out eco balls!

- Use a lower-temperature wash for clothes which aren't very dirty: for most washes, 30°C is just as good as 60°C. The lower the temperature, the less energy used and the lower your electricity bill.

- Use the economy programme where possible.

- If your machine has a cold wash option, try using it for lightly soiled clothing. Most detergents work extremely well at low temperatures.

- Switch the washing machine off at the socket when the programme is finished.

FACT:

THE AVERAGE TUMBLE DRYER USES JUST OVER 4KWH OF ENERGY AND PRODUCES OVER 2KG OF CO_2 EVERY CYCLE. EACH TUMBLE DRYER CONSUMES OVER 1,000KWH OF ELECTRICITY PER YEAR, RELEASING OVER 540KG OF CO_2 INTO THE ATMOSPHERE.

DRYING

Tumble dryers are really energy-hungry machines.

- Air-dry your clothes on clothes racks or lines if possible.

- If you have to use a tumble dryer, then spin dry or wring the clothes before putting them in it. Clean out the 'fluff filter' every time you use the dryer: it will use less power and your clothes will dry quicker.

- Switch the tumble dryer off at the socket when it has finished. It consumes almost 40% of the power whilst on standby.

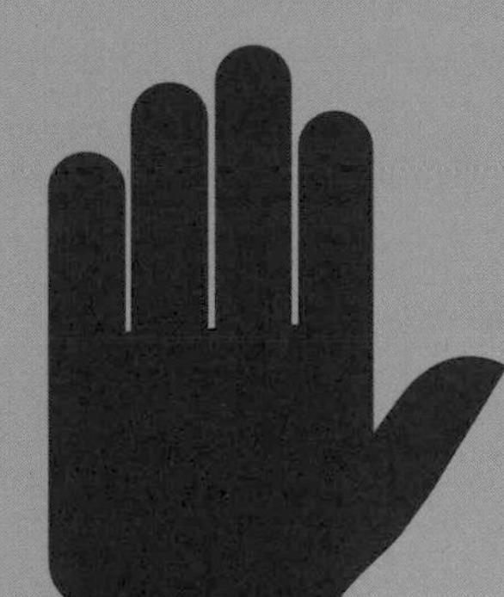

YOU

YOUR BATHROOM

• Don't keep the tap running whilst cleaning your teeth. A running tap can waste up to 10 litres of water in the time it takes to clean your teeth.

• Put the plug in the sink and only run as much water as you need when having a wash – obviously!

• Check all the taps on basins, baths and overflow pipes for leaks or drips – these are usually easy to fix.

• Find and use environmentally friendly toiletries and cosmetics.

FACT:

THE AVERAGE TOILET IS FLUSHED 5,000 TIMES PER YEAR – THAT'S ABOUT 45,000 LITRES OF WATER IF IT'S AN OLDER TOILET.

FACT:

TAKE A SHOWER INSTEAD OF A BATH: YOU'LL ONLY USE A QUARTER OF THE WATER AND ENERGY. (HOWEVER, THIS IS ONLY THE CASE IF THE SHOWER IS NOT A POWER SHOWER – WHICH CAN USE AS MUCH WATER AND ENERGY AS A BATH IN AS LITTLE AS FIVE MINUTES – AND YOU DON'T STAY IN IT FOR AGES.)

YOUR LOO

• Make friends with a hippo! Older toilets can use up to 9 litres of clean water with every flush, while new toilets use only 6 litres. If you're in a house with old toilets, you can reduce the amount of water they use by fitting a water-saving 'hippo' or 'save-a-flush' bag into each cistern. You can get these free from your water company.

IT'S YOUR SHOUT!

If you are in halls with old-style toilets, speak to the warden about getting hippos for them and report any leaky taps.

IF IT'S YELLOW,
LET IT MELLOW.
IF IT'S BROWN,
FLUSH IT DOWN.

YOUR ROOMS
Many cleaning products contain chemicals that have been derived from oil or pollute the planet. Try to use environmentally friendly cleaners, ideally in refillable containers, or make your own and save a fortune – vinegar, lemon juice and bicarbonate of soda can all be used.

www.lowimpact.org
www.stainexpert.co.uk

STOP

DO YOU NEED TO FLUSH?

08
USE
ELECTRONICS

HOW TO USE ELECTRONICS AND NOT DESTROY PLANET EARTH

In the UK our electricity use grows by about 3% a year, so just to keep our CO^2 emissions at existing levels we need to reduce our consumption by 3% a year – and we're told by climate scientists that we need to reduce our emissions by 90%!

FEW STUDENTS ARE without a laptop, iPod or CD player and mobile phone – and that's a lot of kit using electricity. There are some really simple things you can do to use less.

TURN IT OFF

• **That's right off – not just on standby! Never leave your TV, computer, DVD or iPod dock on standby – appliances use a huge amount of power even when they are on standby.** Set-top boxes on standby consume about 85% of the power that they use when they are working, and some use even more than that.

• Most set-top boxes and digital TV recorders can be switched off when not in use, although there are still a few that need to be kept on, including some subscriber services. Check out the instruction manual just to be sure.

• Frustratingly, there are still some appliances being made that have to be kept on, as they need retuning if they are turned off. Digital radios are a prime example – again, check out the instruction manual.

• Switch your mobile phone and laptop chargers off when they are not in use – otherwise they continue to consume electricity. Put your hand on the charger and you will feel all that heat energy just wasting away.

• Use a laptop rather than a desktop computer if you can – they use much less power.

• Set your computer to go on standby if you are going to be away from it for more than five minutes.

STOP

HOW MANY GADGETS HAVE YOU GOT ON **STANDBY** RIGHT NOW?

A journey by air creates about 10 times as much CO_2 as a similar journey by train.

TRAVEL

HOW TO TRAVEL AND NOT DESTROY PLANET EARTH

GET ON YOUR BIKE

- Try cycling for a day a week to start with, and choose days when the weather is fine. You'll save yourself money and become fitter in the process. Break the car habit, reduce your carbon emissions and add years to your life!

HOP ON A BUS

• If you don't have a bike, jump on a bus – they are one of the most energy-efficient ways to travel. And don't forget to check out bus fares when travelling home or visiting friends in other towns and cities – you can get some really cheap deals.
www.megabus.com/uk

LEAVE THE CAR

If you have a car then leave it parked as much as you can; use public transport, walk or cycle instead. Bikes are a great way to get to lectures – they are cheap, carbon-neutral, easy to park, generally quicker than cars in towns and cities, and good exercise. Can you share your car with friends and save CO_2 and some cash?
www.carshare.com
www.liftshare.com

DITCH THE PLANE

Try to avoid flying: air travel is particularly damaging to the environment, adding

much more CO_2 to the atmosphere than any other form of transport. Now we have a high-speed train link to Europe, trains are a fast and environmentally friendly way of travelling; they are as quick as flying in many cases, and drop you in the middle of a city. Don't forget to get a student pass and save on the fare.
www.eurostar.com
www.seat61.com
(this site enables you to plan your travel worldwide using trains, ferries and ships).

TRAVEL TOGETHER

• There are great deals available for people travelling as a group. Many train companies will give you large discounts even if there are only three of you travelling together, and larger groups get larger discounts. Phone them and speak to the group travel office.

• If there is a ski/snowboard trip happening, rather than flying, check out the ski trains: they deliver you directly to the resort – and you get two more days on the slopes.
www.eurostar.com/UK/uk/leisure/destinations/direct_services/ski_train.jsp

Here are some ideas to 'suggest' to your college or university to help staff and students switch to greener ways of travelling.

• Provide easily accessible information about public transport with maps showing cycling and walking routes, timings, timetables and contact names via your green notice board, email, your Local Area Network (LAN) or intranet.

• Approach your local bus company (armed with facts and figures) to see if they will re-route a bus at key times if your college or faculty is not well served by public transport.

• Create good facilities for cyclists: secure bike storage, showers and lockers.

• Set up a 'bike buddy' and/or 'walking buddy' database through email, your Local Area Network (LAN), the university or college website/intranet, or notice board.

• Set up a car-share database.

• Ensure that all visitors are made aware of public transport links and of the university/college policy to reduce car use.

The member of staff in charge of this will have a title such as 'Environmental Coordinator' or 'Energy Advisor'. If your uni or college doesn't have one, check with the Student Union – see if there are some students doing things; if there is nothing happening then it's over to you! Go to http://peopleandplanet.org – for help getting started.

STOP

CAN YOU DITCH THE CAR AND GO BY BIKE?

10

HOW TO SHOP AND NOT DESTROY PLANET EARTH

Your traditional Sunday lunch of roast chicken with all the trimmings could have travelled over 25,000 miles – the chicken from Thailand and the 'fresh' out-of-season vegetables from Africa – a huge number of food miles on your plate!

EVERYTHING YOU BUY, from iPods to bottled water, affects your carbon footprint – the amount of CO_2 your lifestyle generates. Some products have more embodied energy in them than others: some travel thousands of miles to reach our shops; some come from unsustainable sources – such as tropical hardwoods – and some use vast amounts of fertiliser (which are oil-derived) and pesticides in their production, such as cotton. Look out for 'greenwash' – choose genuinely eco-friendly products!

FOOD

• Buy locally grown food, in season, from your local food shops.

• Reduce your food miles – avoid food that has travelled a long way to reach you.

• Eat less meat; the less you eat, the less CO_2 is emitted. Meat production is very energy-intensive.

• Remember: dairy produce is also linked to meat production. Why not cut down when you can?

• Shop online and get your goods delivered.

• Only buy what you need – avoid BOGOFs (Buy One Get One Free) unless you can use the 'free' item or store it.

• Plan your shopping – make a list – and avoid impulse buying and making more trips than necessary.

- Buy organic if possible: organic food and clothing will have been grown without the use of artificial fertilisers and pesticides.

- Keep an eye on what's in the fridge and use it before its sell-by date.

- Use up your leftovers – have a leftovers party! **www.lovefoodhatewaste.com**

- Avoid pre-prepared meals – they are expensive for you and the planet as they are the most frequently thrown-away food item.

- Avoid plastic bags and unnecessary packaging – leave it at the checkout.

- Bottled water mainly comes in plastic bottles, and often from abroad, with a high carbon footprint. Store drinking water in a glass bottle in the fridge – tap water is around 500 times cheaper than bottled water and has a much smaller carbon footprint – leave it overnight and see if you can taste the difference.

CLOTHES

• Borrow/swap rarely used clothes and other items from friends.
www.whatsmineisyours.com

• Use charity shops or eBay.

• Learn to sew!

APPLIANCES & GADGETS

• Get a laptop rather than a desktop computer – they use less energy.

• Avoid plasma screens – they use loads of energy.

• If your gizmos use batteries, recycle them – don't throw them in the bin they will leach toxins in landfill.

• Use rewritable CDs.

• Recycle your old mobile – don't throw it away.
www.mobilephones4charity.com

CUT DOWN

- Avoid using plastic bags – get a jute or hemp bag and reuse it. Why not get a university brand going?
- Buy second-hand and recycled if you can.
- Buy goods that will last.
- Buy less!

GETTING TO THE SHOPS

- Use a bike or the bus for your shopping trips.
- Share a car – shop with a friend.

STOP

CAN YOU BUY LOCALLY GROWN FOOD?

11
THROW
AWAY

HOW TO THROW AWAY AND NOT DESTROY PLANET EARTH

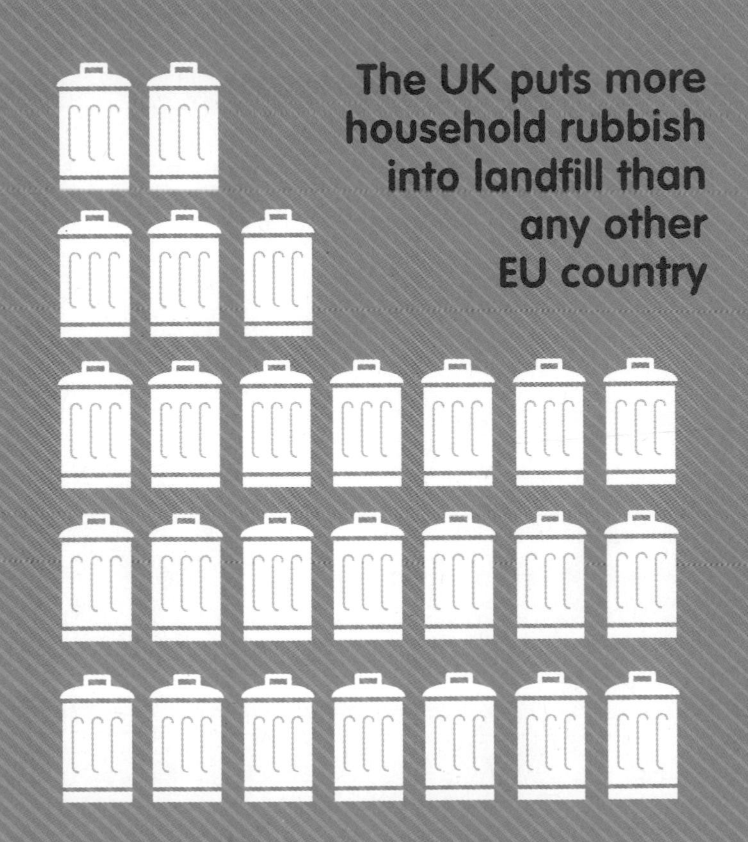

WHAT'S THE PROBLEM?

1. Methane – a powerful greenhouse gas that is released when organic rubbish decomposes in landfill sites.

2. Everything you buy has a carbon footprint – which includes the energy used during manufacture as well as the miles it travels from where it is made to your home.

3. The more things you buy, the more you throw away – and vice versa.

4. We are running out of landfill space.

REUSE AND RECYCLE

• Use your council bins – separate your rubbish into the various different-coloured recycling containers so that as little as possible goes to landfill.

• Use your local recycling facility to recycle anything not collected by your council: plastic, cans, glass, newspaper, wood, telephone directories, electronic goods, cardboard, batteries, electrical appliances and household goods can all be recycled. Ring your local council to find out where the nearest recycling facility is.

• Sell or donate things to a charity rather than throwing them away.
www.freecycle.org

COMPUTERS
Rather than replacing your computer, why not save money and upgrade it instead? If you do buy a new one, give your old computer to a charity. They will wipe the hard drive and donate it to people in developing countries.
www.computeraid.org

MOBILE PHONES
Recycle them – don't send them to landfill. You can get Freepost envelopes from many charities.
www.mobilephones4charity.com

PACKAGING
Avoid pre-packaged food and recycle packaging if you can.

PAPER
Reuse and recycle paper and envelopes and buy recycled where possible.

PLASTIC
Say no to plastic bags: plastic is made from oil, and waste plastic ends up in landfill or pollutes the countryside and our oceans.

FACT:

THERE ARE 300,000 PARTICLES OF PLASTIC RUBBISH PER SQUARE KILOMETRE OF SEA SURFACE.

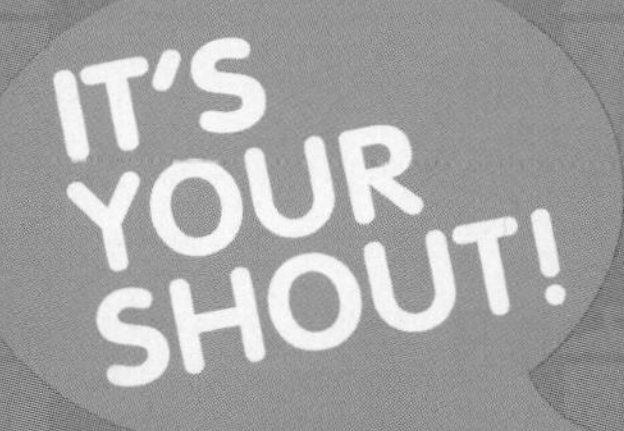

If you are in halls and there are limited recycling facilities, talk to your warden.

STOP

CAN YOU RECYCLE YOUR BEER CANS?

12

SPEND MONEY

HOW TO SPEND MONEY AND NOT DESTROY PLANET EARTH

The annual amount of student loans taken out by new students alone is about £640 million.

WHAT YOU DO with your money affects climate change. This ranges from the glaringly obvious, such as not buying a car with a huge engine that pumps out loads of CO_2, to more indirect actions like placing your student loan in a bank or building society that has a strong climate-change policy and only invests in businesses that are actively working to reduce their carbon emissions.

The majority of high-street banks and building societies only give lip service to addressing the massive problem of climate change. They prefer to ignore the problem and continue to invest your money in companies and businesses that are actively participating in and encouraging the burning of fossil fuels and thereby the release of vast quantities of CO_2.

AS STUDENTS you have considerable purchasing power and can therefore influence the directions of businesses, both large and small. Some companies are still choosing to ignore their impact on climate change, as they feel that if they make changes to reduce their impact it will negatively affect their business. Vote with your money – businesses never ignore that. Shop around: ask the bank what it is doing to mitigate climate change and where it invests your money – and take your money elsewhere if you're not happy with its response. The message will soon get through if enough students move their student loans!

www.triodos.co.uk
www.co-operativebank.co.uk

STOP

WHAT IS YOUR BANK DOING ABOUT CLIMATE CHANGE?

13

WANT
TO DO
MORE?

ACT NOW!

HERE ARE JUST a few ideas for raising awareness of climate change and the need for urgent action.

- **Talk to friends and family** about climate change.

- **Join People and Planet**, the UK student campaigning organisation working to protect the environment, end world poverty and defend human rights. See if your university has an active environmental campaign; if it hasn't, then how about getting one started?

http://peopleandplanet.org

• **Join 'Stop Climate Chaos'**, a coalition of over 100 organisations including Friends of the Earth, Greenpeace, Oxfam, Tearfund, Women's Institute, People and Planet, RSPB and WWF.

www.stopclimatechaos.org

• **Get radical**, join in the various marches – see their website to find out when the next one is – and drag your friends along.

• **Join the UK Youth Climate Coalition** – see what is happening. Get involved.

http://ukycc.org/team/

• **Join 10:10**, and commit to cutting your CO_2 emissions by 10% in 2010.

www.1010uk.org

• **Lobby your university or college** to sign up to 10:10. Use this to generate a powerful PR message about your university or college. Tell your local radio and TV stations about it.

• **Use your influence** – ask your MP to join the 10:10 campaign. Send him or her a letter or email; make yourself heard – it's your future. (To find out who your MP is see **www.theyworkforyou.com.**)

www.38degrees.org.uk/page/speakout/1010

• **Organise a showing of An Inconvenient Truth** or **The Age of Stupid** – help other students understand the urgency of the problem.

www.climatecrisis.net/aboutthedvd/index.html

www.ageofstupid.net

• **Download posters and stickers** to use as part of your campaigns from

www.carbontrust.co.uk.

• **WISE UP** – read all about it.
Here are a few books to get you going:

Carbon Counter.
Mark Lynas, Collins, 2007.

Field Notes from a Catastrophe: A Frontline Report on Climate Change
Elizabeth Kolbert, Bloomsbury, 2007.

Global Warning: The Last Chance for Change.
Paul Brown, A&C Black, 2006.

Heat: How We Can Stop the Planet Burning.
George Monbiot, Penguin, 2007.

Sea Change: Britain's Coastal Catastrophe.
Richard Girling, Eden Project Books, 2008.

Six Degrees: Our Future on a Hotter Planet. Mark Lynas, Harper Perennial, 2008

GO!
WHAT CAN YOU DO?

Other Green Books to help you cut your carbon:

Reduce, Reuse Recycle
by Nicky Scott

"Packed with superb information on getting the most from what we use" – www.naturalchoices.com

This easy-to-use guide has the answer from all your recycling questions. Use its A–Z listing of everyday household items to see how you can recycle most of your unwanted things, do your bit for the planet, and maybe make a bit of money while you're at it.

ISBN 978 1 903998 93 9 £4.95 pb

The Use-It-All Cookbook
by Bish Muir

Loads of ideas for using leftovers, including that sad-looking carrot or hard cheese at the back of the fridge. Soups, stews, pies and risottos sit alongside delicious quick recipes and tasty juices and smoothies.

ISBN 978 1 900322 30 0 £12.95 pb

Other Green Books to help you cut your carbon:

Cycling to Work
by Rory McMullan

"A handy guide to get you cycling to work – and once you start, you won't want to stop!" – Sustrans

This book is packed with great tips on: buying the right bike and equipment, riding safely in traffic, finding the best route to work, integrating cycling with other forms of transport, basic bicycle mechanics, and finding bike buddies.

ISBN 978 1 900322 12 6 £4.95 pb

Cutting Your Car Use
by Anna Semlyen

Whether you go the whole hog and sell your car, or just reduce your car use, this book will help you to a fitter, more sustainable lifestyle.

ISBN 978 1 900322 15.7 £4.95 pb

Other Green Books to help you cut your carbon:

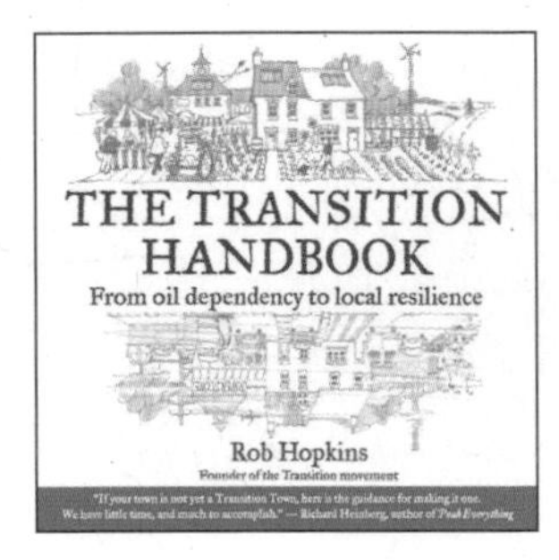

The Transition Handbook
by Rob Hopkins

"The Transition movement is the best news there's been for a long time, and this manual is a goldmine of inspiration to get you started." – Phil England, New Internationalist

This manual of the fast-growing worldwide Transition movement shows how communities can deal with the twin challenges of peak oil and climate change, and is a guide to beginning your 'energy descent' journey.

ISBN 978 1 900322 18 8 £12.95 pb

Your Money: use less, save more
by Jon Clift and Amanda Cuthbert

There are lots of things you can do to save money – and a lot of money-saving ideas give you the chance to get together with other people, cut your waste and slow down a bit. This A–Z guide is packed with tips on saving money on everything from credit cards to car boot sales, from toys to travel.

ISBN 978 1 900322 53 9 £4.95 pb